THE WORLD'S EASIEST ASTRONOMY BOOK

Hitoshi Nakagawa

ONE PEACE BOOKS

Thoughts for Space: Preface

On my first spring break as a university student, I went to the NASA Kennedy Space Center in Florida. It was my first ever trip abroad; I was going to see the Space Shuttle launch.

The night before the launch, the Shuttle was revealed, lit up on the launch pad. The towering 184-foot rocket standing before me was about the same height as a 15-story building. The Shuttle is the world's first winged re-entry flight space vehicle. I was overwhelmed by its beauty and was almost moved to tears. The Shuttle's white, smooth form looked like a white bird. Yet, this was no ornament or work of art, but a vehicle to carry people into space. I was amazed that a combination of all the latest developments in space engineering could be this beautiful.

Finally, the morning of the launch dawned. A countdown ticked away on a giant digital board, and at the moment of the launch everything fell silent. Seconds later came a sound like an explosion: a tremendous roar which vibrated throughout my body. A dazzling flame flared out of the main engine, white smoke gushed out in ballooning clouds, and the white bird flew away from the Florida plains, picked up the Atlantic winds, and was swallowed by the clouds. After the white bird had flown, all that remained was a towering pillar of smoke rising from the ground into the clouds, like a stairway to the heavens. At that moment I felt the nostalgic happiness of my childhood, a scene I will never forget.

Now, in the 21st Century, mankind is just beginning to expand the realm of life, from on the Earth's surface to in space. Following on the Space Shuttle flight and the International Space Station, manned flight to the Moon will soon resume. After that, a mission to Mars will be launched from a Moon Base. I wonder how far mankind will reach in this century. The mysteries of the universe are now gradually being solved using new technologies such as the Hubble Space Telescope and planetary exploration. It would be wrong to leave all of this to astronomers only. To add to our knowledge of the universe should be a joy which uplifts the hearts of all mankind.

I have written this book for those who have no in-depth knowledge of space: I intend to offer them a feel for the mysteries, wonders, and marvels of the universe, and to provide a taste of mankind's efforts in space. The content is based on answers to simple questions from general readers with no particular connection to astronomy.

When I gaze at the stars twinkling in the night sky, my thoughts turn to the endless universe. The light from those stars is likely to have started its journey 2 million years ago. Humankind seems insignificant compared to the vastness of the universe, yet we insignificant humans have built a space station that orbits the Earth and houses a crew even now. Looking up from the Earth, it is possible to see the space station, twinkling away just like one of the many stars. In our busy day-to-day lives we often forget to lift our heads and look to the skies, which is exactly why we must look up and open our minds to the universe. In the past, in the present and onwards to the future, space and time are endless. I want us all to surround ourselves with thoughts of mankind's advance into the universe.

Just as I was deeply moved by the Space Shuttle when I was young, which led me to a life in astronomy, I hope that this book will spark your imagination and ideas about the universe, convey the excitement of setting out on the Age of Discovery of space, and bring joy to the reader.

Hitoshi Nakagawa

Osaka Prefecture High School Teacher

Former Japan Aerospace Exploration Agency (JAXA) Officer

THE WORLD'S EASIEST ASTRONOMY BOOK

CONTENTS

Lesson01

Do Aliens Really Exist?

The simple answer is that nobody has found one yet. Any stories you might have heard about aliens visiting Earth in UFOs, or being hidden away or captured by NASA are almost certainly not true. As far as we know, no aliens have visited Earth and there is no evidence that they exist at all. In fact, no living things have been discovered anywhere outside the Earth, not even tiny microbial life.

Having said that, there is also no proof that aliens don't exist!

There is one group of astronomers, called SETI*, who take the search for alien life seriously. They use huge radio dish telescopes to send out human messages to alien life and scan the skies for radio waves which might contain messages from alien species. The head person at SETI, an American astronomer, Dr. Frank Drake, proposed an equation to estimate the number of civilized stars in the Milky Way. We call this the Drake Equation. Opinions differ on how the Drake Equation should be interpreted, but some scientists estimate that there could be more than one million stars with intelligent life in our galaxy alone.

Way back in the 1970s, four space probes were launched called Pioneer 10 and 11 and Voyager 1 and 2. Even now, they are still on route out of the Solar System. Each probe carries a message to aliens who might stumble across them in the future. Aluminum plaques on Pioneer 10 and 11 show the location of the Earth and an image of the human form. Voyager 1 and 2 contain gold-plated copper disks with audio recordings of music, various nature sounds, and greetings in 55 languages (even konnichiwa in Japanese) as well as the Earth's images. A long journey into the deep space has just started. It could take as much as two to four million years before they reach our neighboring stars.

Later, in 1996, NASA grandly announced that they had found traces of tiny microbial life in a meteorite from Mars. Unfortunately, however, further analysis showed the evidence to be quite weak.

Then in 2004, NASA sent out two robotic vehicles–Mars Rovers–to the surface of Mars coinciding with the closest approach of Mars for 60,000 years. The rovers discovered that there had been lots of water on Mars in the past. This was an amazing discovery, because if there was water then it's possible there was life too. If proof was found that life of any kind exists (or existed) outside the Earth that would mean there is a much greater chance of there being intelligent life somewhere in the vast universe.

So, could we humans be the only intelligent life in the universe? Or, is the universe teeming with intelligent life that we have yet to discover? One thing is for sure – the quest to find alien life is not over yet.

We are knocking on the door of the universe,
Calling out to our friends we have not yet met.

* SETI: Search for Extraterrestrial Intelligence

Lesson 02

How Many Dimensions Are There?

When you watch a firework go bang in the night sky, the explosion looks like a circle no matter which direction you're looking from. This is because the explosion is a spherical shape. This is an example of how you can think about something in either 2 dimensions or 3 dimensions. It helps us to visualize things in a different way.

A simple way to visualize the universe is as a 3 dimensional space with 3 axes which meet at right angles. You know these axes as length, depth, and height. Scientists often visualize a 4 dimensional universe using time as the fourth axis, which is why you might have heard the term space-time.

Imagine for a moment an ant walking on the surface of the Earth. It would only be able to imagine walking forwards, backwards, left, and right, and would visualize its universe as an infinite flat surface. In reality, we know the world is a 3 dimensional space because the Earth is curved, and its surface is not infinite. We can visualize this because we can look up to the skies. If we were only to look in front of us, all we would see is the 2 dimensional space of the ant.

But what if, just like the ant, we too are unable to see a further dimension? Perhaps we only see 3 dimensions even though we live in a 4 dimensional space. Such a 4 dimensional space would have length, depth, height, and one more dimension, all meeting at right angles. For us humans such a space is impossible to imagine.

However, we are able to describe such a space, or even a 5 or 10 dimensional space, using mathematics. So, perhaps there is room yet for further leaps in the imagination and ingenuity of humankind.

We cannot easily imagine it because we have yet to experience it.

0 Dimension: Point
1 Dimension: Line
2 Dimensions: Surface (Length x Depth)
3 Dimensions: Space (Length x Depth x Height)

Lesson 03

Everything We See is In the Past

The speed of light is 300,000 km (186,000 miles) per second, meaning that light could circle the Earth seven and a half times in a single second. Even at this incredible speed it still takes light from the Sun 8 minutes to reach the Earth. That means that when we see the Sun, what we actually see is the Sun from 8 minutes ago. Even if the Sun were to explode we wouldn't know it until 8 minutes later. In the same way, when we look at a star 100 light years* away with a telescope, we are seeing what that star was like 100 years ago.

In 2003, the age of the universe was calculated to be 13.7 billion years based on observations by NASA's Microwave telescope WMAP**. That means that the Big Bang, the event which started the universe, took place 13.7 billion years ago, and the universe has been continuously expanding at the speed of light ever since. You could also describe this by saying the distance to the farthest edge of the currently observable universe is 13.7 billion light years. Using this number we can calculate the size of the universe to be 13.7 billion multiplied by 9.46 trillion km (1 light year). This is the size of the universe. Isn't it simply impossible to comprehend?

One implication of this is that we can't see anything more distant than 13.7 billion light years away. That's because nothing existed before the Big Bang. This edge is called the horizon of the universe. By observing the limits of the universe in this way, we can begin to understand its origins.

This effect is due to the speed of light, and it doesn't only affect stars and galaxies. Just looking at the view outside your window or at a calendar hanging on your wall, what you are seeing is always an image from the past. Light is reflected by objects, and shortly afterwards arrives at your eyes allowing you to see it. The same is true even for this book. Hold this book 12 inches from your eyes and you're looking at an image of this book 0.000 000 001 seconds in the past.

Look to the limits, understand the origins.

* 1 light year = the distance that light travels in 1 year = 9.46 trillion km (about 5.88 trillion miles)
** WMAP: Wilkinson Microwave Anisotropy Probe

Lesson 04

Everything is Pulled Together

Legend has it that in the 17th century, Newton watched an apple fall from a tree and discovered universal gravitation. Universal gravitation is a fundamental law which applies to all things in the universe and is a force of attraction between one object and another. Universal gravitation acts between the Moon and the Earth, between a pencil and an eraser, and even between this book and you.

From our standpoint it appears as though all objects are pulled towards the Earth (downwards). It's only because the attraction between the Earth and objects is so overwhelmingly powerful compared to the attraction between other objects. We call this attraction between the Earth and objects gravity.

Gravity (universal gravitation) is stronger the closer you are to the Earth, and weaker the further away you go. So, why is it that people and objects inside a spacecraft float around? Is it because they are so far away from the Earth there is no gravity?

That's not quite right. Even if the spacecraft was 400 km (250 miles) high in outer space, there would still be an attractive force towards the Earth of around 90% of that at the surface. If an object was just simply placed up there it would still fall down towards the Earth. The reason that people and objects inside a spacecraft are able to float is that the spacecraft is actually continually falling towards the Earth. Although normally objects that fall will eventually land, orbiting spacecraft counteract this by flying forwards at fantastic speeds producing a constant circular motion around the Earth. It's this state of free-fall that creates a constant effect of zero gravity.

To imagine this in a simple way, imagine you are riding in an elevator. When it suddenly descends, you get a floating feeling in your stomach. If the wire were suddenly to be cut, the inside of the elevator would be in a zero gravity state (until it crashes to the floor!). You can also experience zero gravity during the fall when bungee-jumping, or a vertical-drop ride at a theme park.

The space station orbiting the Earth is like an elevator which is forever falling, and the inside of it is a stable zero gravity state (or more accurately a micro-gravity state). Such a unique environment allows cutting-edge research to be carried out on the development of new materials, biology, and medicine.

The Earth's gravity reaches into space.
Fall freely in gravity and you'll feel zero gravity.

Lesson 05

Living at Light Speed

The speed of light is 300,000 km (186,000 miles) per second.

If there was a vehicle such as a spaceship that could fly close to the speed of light and you took a trip on it mysterious phenomena would occur. As you approached the speed of light the length of the spaceship would get shorter, the mass of the spaceship would increase, and time inside the spaceship would move forward more slowly. Not only that but the closer you got to the speed of light, the heavier the spaceship would get, and you would never be able to reach the speed of light.

Einstein's special theory of relativity says that light speed is the only absolute in the universe, and other things like length, mass, or time change depending on the observer and the conditions. According to the special theory of relativity, nothing can travel faster than the speed of light.

But even without exceeding it, as you approach light speed you experience relative time. If you were to take a one-year (365 day) trip on a spaceship traveling at half light speed and then return to the Earth, 421 days would have passed, effectively meaning you would travel forwards in time by 56 days and arrive back at the Earth in the future. If you were to take a one-year trip on a spaceship traveling at 99% light speed and then return to Earth, 7 years would have passed on the Earth and your friends would have aged by 6 more years than you. If you were to take a one-year trip on a spaceship traveling at 99.999% light speed and then return to Earth, a staggering 224 years would have passed on the Earth!

Using the special theory of relativity like this, it is possible to travel to the future. However, with today's technology traveling at half light speed is just a dream. We don't even have a rocket which can travel at 1% of light speed. The speed of the space station or the Space Shuttle is somewhere around 0.003% of light speed. The special relativity time-slip when you take a one-year trip at that speed is only around 0.01 seconds.

Only the speed of light is absolute.

Lesson 06

A One-Way Ticket to the Future

The Pole Star, landmark of the northern skies, is 430 light years from the Earth. The light you can see coming from the Pole Star was emitted around in Colonial times. It would take 430 years to travel from the Earth to the Pole Star, even traveling at the speed of light (as viewed from the Earth).

We know that nothing can travel faster than light, so you would imagine that it's impossible to reach the distant Pole Star in a single person's lifetime. However, the special theory of relativity tells us that the flow of time is not fixed, but differs depending on your location (depending on the observer). If a spaceship was developed that could fly at 99.9999% of the speed of light and you took a ride in it, the flow of time would be 1/707 of that on the Earth. If you were to travel to the Pole Star at that speed, it would take no more than around 7 months. To the people inside the spaceship, it would seem as though they had broken the speed of light and traveled 430 light years in 7 months. On the Earth, though, the full 430 years would have passed.

By riding in spaceships close to the speed of light it is possible to travel tens of thousands of light years, or tens of thousands of years into the future. Beware, though, because there is no way to go back to your own time. It's only a one-way ticket to the future.

If it were possible, would you go?

The flow of time depends on your point of reference.

Lesson 07

When the Moon Looks Large

Have you ever noticed the Moon looking larger than usual, as though it were closer to the Earth than normal? This is only an optical illusion.

The orbit of the Moon is an ellipse, so the Moon does move ever so slightly closer and further away from the Earth as it revolves. However this difference is very slight and hardly affects the appearance of the Moon at all.

When the Moon appears larger than normal, it is in a low position close to the horizon. To put it another way, when the Moon is low you can then compare it to tall objects such as buildings, mountains, and trees. The Moon, being very far away, is compared to things close-by, so the "large Moon phenomenon" occurs.

For those who don't believe it, hold up your thumb at arm's length around the time of a Full Moon, and compare the size of the moon to your thumbnail. When you try it, whether you think the Moon looks larger or smaller, you'll see that the Moon has not actually changed in size.

The Moon appears larger when there are things close-by for comparison.

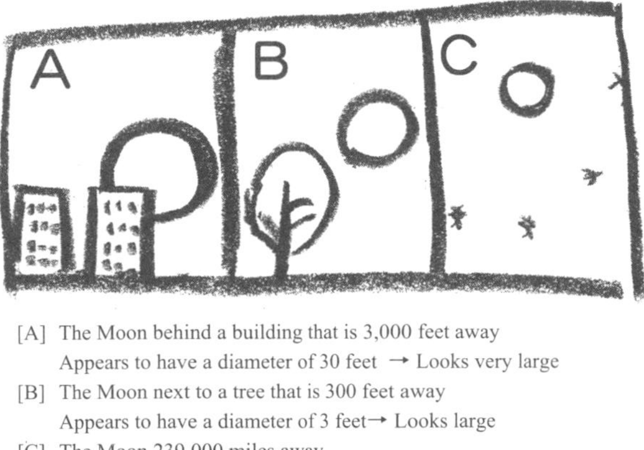

[A] The Moon behind a building that is 3,000 feet away
Appears to have a diameter of 30 feet → Looks very large

[B] The Moon next to a tree that is 300 feet away
Appears to have a diameter of 3 feet→ Looks large

[C] The Moon 239,000 miles away
A diameter of 2,200 miles = Actual size

In each of A, B and C, the Moon will be comparably the same size when held up against your thumbnail 2 feet from the eye.

Lesson 08

Life of the Universe

It is said that all things which have a beginning must also have an end. Fashions eventually fade, buildings eventually crumble, and living things will eventually die. Even a star has a lifespan. Our Sun is currently 4.6 billion years old, but its lifespan is 10.9 billion years. Scientists believe it will cease to shine in around 6.3 billion years.

The birth of the universe was 13.7 billion years ago. That birth was a huge explosion called the Big Bang, and the universe has been continually expanding like a balloon from then right up to the present day. But does the universe have an end?

Well, there have been a number of theories on that. There are two major theories. On the one hand, there is the theory that the universe will continue expanding forever, and on the other there is the theory that once the universe has expanded to a certain point it will gradually collapse again, right down to a point and then it will disappear.

An answer to this question was finally provided in 2003. The mass of the universe was measured by NASA's WMAP microwave orbiting probe, calculating that the universe will expand forever.

The only thing we can say forever is the universe.

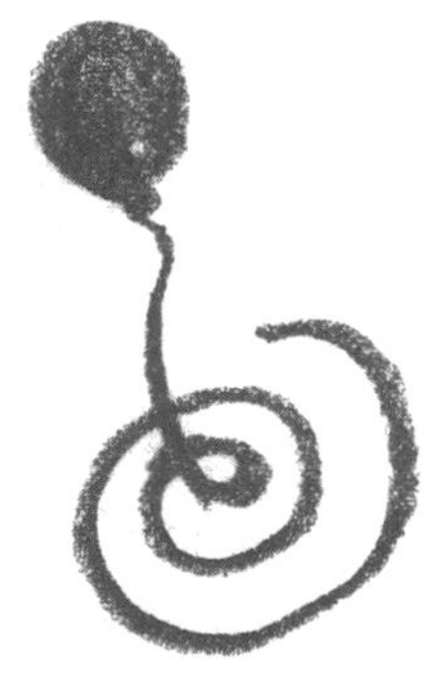

Lesson09

SPACE ELEVATOR

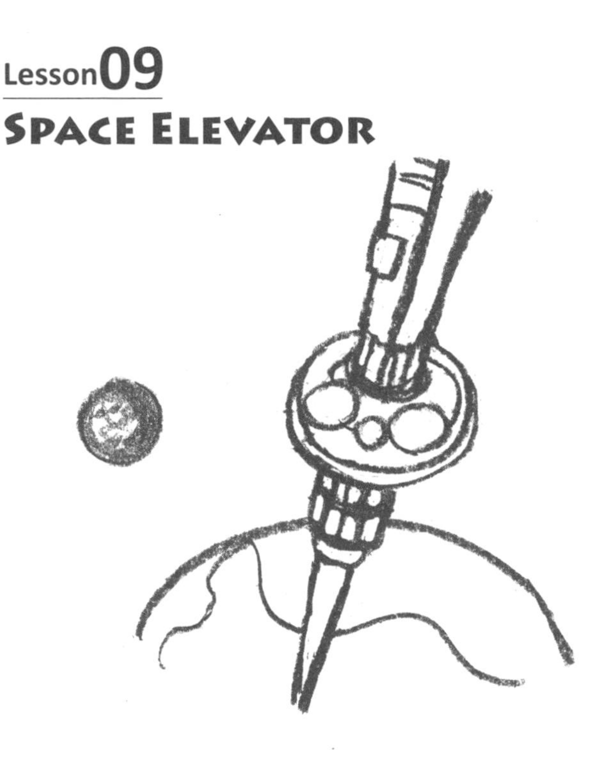

In the novel *The Fountains of Paradise,* written in 1979 by Arthur C. Clarke, a space elevator (or orbital tower) was described.

The space elevator is a concept to solve the problem of how to travel to and from space, by constructing a giant elevator in a long tube connecting the surface of the Earth to space. If such a device were to be made it would be possible to travel between the Earth and space safely, at low cost, and with low pollution.

You could think of it as a tube hanging from space down to the Earth. Alternatively, you might think of it as a giant tower rising from the Earth all the way into space. A little like the beanstalk that climbed above the clouds in Jack and the Beanstalk.

You could ride up the elevator to an orbiting geostationary satellite 36,000 km (22,000 miles) high. Geostationary satellites look like tiny seemingly static dots in the sky when viewed from the Earth's surface. This is because they complete one orbit around the Earth's equator every 24 hours (weather and television broadcast satellites are examples of geostationary satellites).

A space elevator would be built from both the Earth and space ends at the same time. It must be built with great care, so the elevator's center of gravity doesn't slip away from the geostationary orbit*. A tube would be extended outward from the geostationary satellite and a weight attached, to balance centrifugal force and gravity. In practice, these centrifugal and gravitational forces would put an enormous tension on the tube producing all sorts of problems in finding suitable materials and structures to withstand them. Another practical difficulty would be the enormous cost. Suffice to say, there are no current plans for such a project. You can find an artist's impression of the space elevator on the NASA homepage, and scientists are looking into whether the idea is theoretically possible.

Early science fiction writers, such as Jules Verne and H. G. Wells, were the first to dream about spaceships, submarines, and computers, as much as 100 years ago. These things are now reality. Perhaps the day will come when the space elevator dreamt up by Arthur C. Clarke will also be a reality. For all new technology, there must first be someone who imagined it.

If it can be imagined, it can be made.

* Geostationary orbit: the orbit at which a geostationary satellite flies (36,000 km or 22,000 miles above the equator).

Lesson 10

Where Does Space Start?

We know that birds and airplanes fly through the sky, while spaceships and space stations fly in space. But where exactly does the sky end and space begin?

In aviation law, there is a boundary between space and the Earth. Usually, the definition from the Federation Aeronautic International is used, which states that space is an altitude of 100 km (62 miles) and above*. A person who has flown higher than an altitude of 100 km has officially made a space flight. The altitude of 100 km is an imaginary boundary in the upper atmosphere chosen simply for convenience. The atmosphere there is very thin and has very little influence. The density of the atmosphere gradually drops as you get higher, so there is still some residual atmosphere above 100 km. Atmospheric science doesn't consider there to be a defined boundary between the Earth and space. The Outer Space Treaty of 1967 states that no nation may claim ownership of outer space. That means the space above 100 km high belongs to nobody and is shared by all mankind as a whole.

In June 2004, the world's first civilian space flight using a private rocket was launched. That rocket was called SpaceShipOne and was developed by an American company called Scaled Composites. On that first flight, the rocket reached 100.12 km, peeking just 120 meters (400 feet) over the boundary into space, before returning to Earth. The trip took around 1 hour from takeoff to landing, and a state of weightlessness was experienced for 3 minutes. This was considered a space flight. Later that year, in October, SpaceShipOne achieved two consecutive space flights in a 2 week period. For the first time, SpaceShipOne gives ordinary people the opportunity to travel into space.

The area beyond the skies belongs to all of us.

* There is also a definition used by the US Air Force of an altitude of 50 miles (80 km) and above.

Lesson 11

Why Is the Sky Blue?

When you look up to the sky on a sunny and cloudless day, the sky is a deep blue. But have you ever wondered why the sky is blue? The secret is in an effect called scattering.

Scattering is a phenomenon where a wave such as light or a radio wave hits an object and is spread from the center in all directions. Light from the Sun is scattered by nitrogen and oxygen molecules and tiny particles in the atmosphere.

The color of sunlight is made up of the 7 familiar colors: red, orange, yellow, green, blue, indigo, and violet, in order of wavelength. Fine particles like nitrogen and oxygen molecules have the characteristic that they scatter bluish light (short wavelength) more easily than reddish light (long wavelength). The reason for the blue sky at daytime is that we see this scattered light.

Similarly, the reason the sunset is red is related to the angle of the

Sun. The atmosphere covers the Earth in an even thickness, so it is thinnest when you look directly upwards and thickest when you look towards the horizon. Light from the setting Sun enters the atmosphere diagonally, so it has to pass through a lot more of it before it reaches us. By the time it reaches the surface, almost all of the bluish light has been scattered away and all that remains is the reddish light.

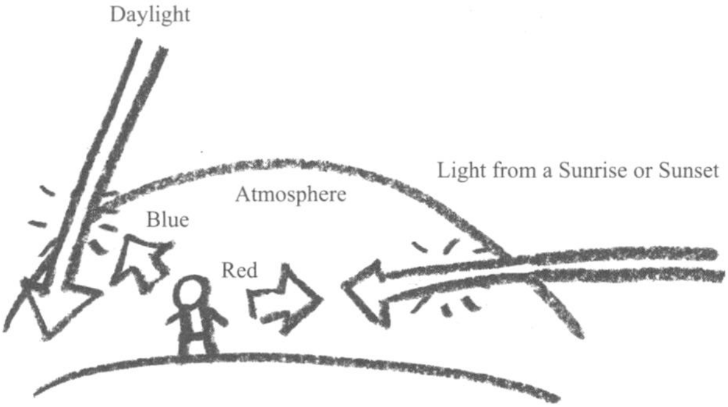

Daylight

Light from a Sunrise or Sunset

Atmosphere

Blue

Red

So what about other planets? Mars, for example, has a low atmospheric pressure and only traces of nitrogen and oxygen in its atmosphere. Gravity there is quite weak and dust particles are easily whisked up into the air. Iron dioxide contained in the Martian dust absorbs blue light more effectively than red light. Therefore, if you looked up the Martian sky during the day, you would not see a blue sky but see a sky with a pink tinge. Larger particles contained in the Martian dust scatters some of the blue light into the area just around the Sun. That is the reason why you would see bluish light near the Sun in a Martian sunset. So, on Mars the daytime sky is pink and the sunset blue.

On bodies like the Moon that have no atmosphere, sunlight is not scattered at all. From the surface of the Moon in daytime you would see a stark white Sun shining amidst a jet black sky.

Clear blue skies and red sunsets go hand in hand.

Lesson 12

Define the Universe

What exactly do we mean when we talk about the universe?

Well, the word universe comes from the Latin word universum, first used by a Roman poet in the first century BC, and means everything combined into one. This term is used to mean all things across all of space and all of time, and so everyone, even you, is a part of the universe. The word cosmos is used to describe the universe as an ordered system as opposed to chaos. The word Space (or outer space) refers to the space outside the Earth's atmosphere. We use this to describe the wide expanse beyond the skies, in comparison to the Earth and its surface.

The science of the origins and the fate of the universe uses the term universe (or cosmos), whereas the science of satellites and space exploration uses the term space.

The universe is full of mysteries. Mankind's discoveries so far are no more than a small fraction of these. Research into the structure and origins of the universe is deeply linked to a deeper understanding of the question of where we came from and where we are going.

The chance of meeting a special person in this vast universe is tiny.

Lesson 13

The Centrifugal Force of the Earth

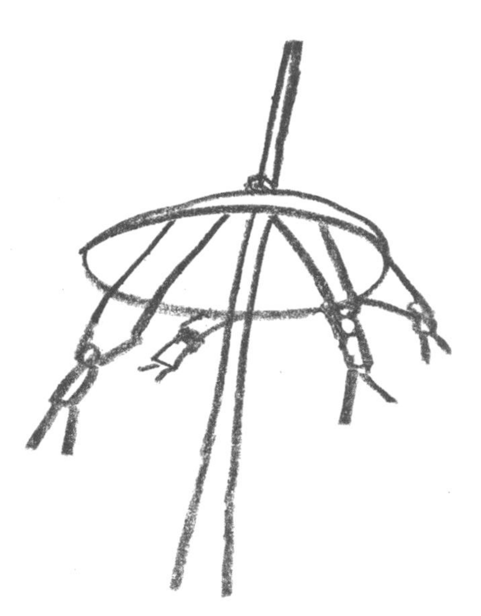

Have you ever been on a self-propelled merry-go-round? It has a number of metal bars coming out from a central pillar which you hold onto as it spins around. You may have seen one in your local playground. When you're spinning on a merry-go-round your body is pulled outwards. The faster it spins the greater the force pulling you outwards, and you feel as though it's throwing you off. The force you feel on the merry-go-round is called centrifugal force. Centrifugal force can be calculated from the angular velocity (the angle moved through in 1 second), and the radius of the rotation. The faster the angular velocity, the greater the centrifugal force. In the same angular velocity, the longer the radius of rotation, the greater the centrifugal force.

The Earth rotates at a rate of 1 rotation per day, or every 24 hours, so someone standing at the equator would feel the greatest centrifugal force because the equator is the point with the longest radius. In Japan (or the United States and other mid-latitude countries), we feel a medium centrifugal force. Someone standing at the North or South Pole would feel no centrifugal force at all. A person standing at the equator is rotating at a speed of 470 meters (1,500 feet) per second (that's faster than the speed of sound!), and a person in Japan (or the U.S.) at around 300 meters (1,000 feet) per second. Although it is an incredibly fast speed, the force of gravity pulling people down towards the Earth is stronger, which is why we don't fly off. At the equator the force of gravity is 290 times stronger than centrifugal force, and in Japan (or the U.S.) it is 440 times stronger.

You might ask what if the Earth was to spin faster? Would people really fly off into space? Well, if the Earth didn't spin at

1 rotation every 24 hours, but at 1 rotation every 84 minutes for example, a person standing at the equator would be traveling at 7.9 kilometers (4.9 miles) per second, which is where gravity and centrifugal force are exactly equal. If it were to spin any faster than this, the person would be thrown upwards away from the Earth by centrifugal force. This speed of 7.9 km/s is the base speed of space flight, called the "first cosmic velocity," and is close to the speed of satellites orbiting near the Earth's surface.

You are moving at 1,000 feet per second right now.

Lesson 14

The Beginning of the Universe

In 1929, an American astronomer called Edwin Hubble observed distant galaxies and noticed that all stars and galaxies were moving away from us. He realized that the universe was gradually expanding. If you were to reverse the speed of this expansion, the universe must have been smaller in the past. Not only that, but if you go back 13.7 billion years the universe would have been a single point.

At that time, 13.7 billion years ago, a large explosion (expansion) took place called the Big Bang which started the universe, and the universe has been freely expanding since then. People call the Big Bang the beginning of the universe, but a recent theory suggests the universe was born slightly before the Big Bang. Ten to the negative 44th power seconds after the birth of the universe, the size of the universe was 10 to the negative 34th power cm. Then, it suddenly expanded to 1cm in an instant. This sudden expansion is called Cosmic Inflation. Inflation was all over just 10 to the negative 33rd power seconds after the birth of the universe. (We call the fireball state just after Cosmic Inflation the Big Bang. The Big Bang was caused by Cosmic Inflation.)

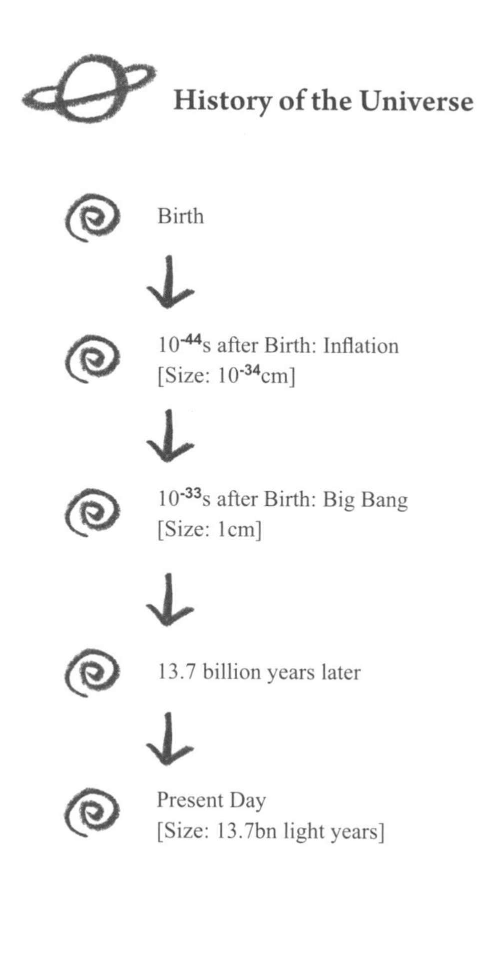

The phrase 10 to the negative number is a mathematical term which means 1 divided by 10 that number of times, so 10 to the power -4 is 1 divided by 10, 4 times, or 0.0001.

A size of 10 to the negative 34th power cm is far, far smaller even than an atom and an electron. Everything that exists in the universe today was crammed into that tiny space. That's a really ultra-high-density state. As well as being the beginning of all matter, the start of the universe is also the beginning of all space and all time. That means it makes no sense to ask "What was there before the start of the universe?" The universe did not start in the midst of the flow of time, the flow of time itself started with the universe. There can't be such a thing as a time before time began*.

The universe began with sudden expansion (Cosmic Inflation).

* A recent theory by the British physicist Stephen Hawking suggests that before the universe began (before our time started) there was imaginary time.

Lesson 15

What Is Outside the Universe?

The universe does not exist inside some space, it is space itself.

If the universe were inside some space, we could ask "What exists outside the universe?" But space and the universe are one and the same. The space we call the universe was created in the Big Bang and then expanded. Space itself expanded.

There is no concept of outside the universe.

Lesson 16

Could Mankind Live on the Moon?

Mankind landed on the Moon between 1969 and 1972. Six spacecrafts, Apollo 11 to 17, have touched down (except for Apollo 13 which returned without landing on the Moon), and a total of 12 astronauts have set foot on the Moon. These 12 astronauts (all Americans) are the only people who have walked on the Moon, and nobody has visited the Moon since 1972.

Now it seems a little like a fairy tale. Science has developed so much, so why did nobody else go to the Moon? Perhaps it really is just a story?

If we take a look through history, we'll see that this is normal. Take Columbus, for example, who discovered the new continent (the Americas) in 1492. Europeans didn't begin to colonize the land, which is now the USA, until 1600, more than 100 years later. Or take Amundsen, who was the first to reach the South Pole in 1911. It was after 1956 that the Amundsen Scott Base (America) and the Showa Base (Japan) were permanently established in Antarctica, 45 or more years later. In both cases, the main focus was on an individual reaching the destination, but it took some time before the new environments could be utilized to enable people to live there.

It's the same with the Apollo missions. Throughout history, enormous effort and money was expended in the effort to reach the Moon. But an entirely new effort would be needed to enable people to live there actually.

However, recently NASA has finally made plans to send astronauts once more to the Moon sometime in the 2020's. If successful, they would be the first men on the Moon in about a half century. After that, plans are being developed for the construction of a base on the Moon.

On the Moon there is no air, and cosmic radiation constantly rains down, so a facility similar to a space station would have to be built under the surface to live in. You can't walk around on the surface of the Moon without a spacesuit, but since the gravity is one sixth of that on the Earth, it would be a very pleasant experience. Your body would feel light, and you would be able to lift heavy objects easily. Records for track and field events like the long jump or hammer throw would be smashed. They may have to come up with a double standard, such as Earth Long Jump 26 ft, Moon Long Jump 156 ft.

We are now living in the interval between the era of reaching the Moon and the era of re-visiting and making use of the Moon. The International Space Station is orbiting above our heads even today, and astronauts live there permanently. Astronauts have stayed in the Russian space station, MIR, for periods of over a year. Technologies being tried out on space stations will one day be used on the Moon, and the day when someone will stay permanently on the Moon is not too distant. At the moment, America, Russia, Europe, China, and Japan are all turning their attention towards Moon probes. The time of a return to the Moon is coming.

Mankind will visit the Moon again.

Lesson 17

Day and Night in Space

The Earth rotates once every 24 hours or once per day. This is what gives us the length of our days and nights, around 12 hours each (changing depending on season and location).

Other planets also have days and nights. Mars rotates once every 24 hours 37 minutes, so the days and nights there are roughly the same as ours. However, the Moon takes 27 days to rotate just once, so day and night last for 2 weeks each! Planets which rotate even slower such as Mercury and Venus can go for 2 or 3 months with neither a sunrise nor a sunset! You could say that places lit up by the sun are called day, and areas in shadow are called night.

But what about the International Space Station which orbits all the way around the Earth in just 90 minutes? The astronauts living in the station have a bewildering blur of days and nights. Day and night are only 45 minutes each, meaning there are 16 sunrises and 16 sunsets in every single 24 hour period. It would be impossible to go to bed and get up again at that sort of speed. That's why life onboard the International Space Station follows GMT (Greenwich Mean Time), which is set on the Earth. The astronauts keep to a schedule by rising at 6 am GMT, and sleeping at 9 pm GMT. Nevertheless, while they sleep the Sun rises and sinks over and over again. The astronauts use sleeping bags and eye masks so they can sleep peacefully.

The cycle of day and night is different for each planet.

Lesson 18

Rotation of the Earth

Unless a force is applied, a static body will remain static and a moving body will continue moving.

This is also true for rotating bodies. A rotating body tends to want to continue to rotate. But as we all know, a spinning top will eventually come to a stop. This is due to friction with the floor. Friction is a force which acts in the opposite direction to motion, or a force which prevents motion. The spinning top feels a force in the opposite direction to its spin, gradually loses speed, and before long comes to a rest.

But, what if we spin a top inside a zero-gravity space station? A spinning top floating in zero gravity would spin well. But although it would spin better than on a surface, it would still slow down. There is no friction with the floor as it's spinning in air, but the friction with the air in the space station would also apply a force preventing motion.

So, bearing that in mind, what about the rotation of the Earth? The Earth is in the vacuum of space and is rotating without touching anything, so you would expect it to always rotate at the same speed. Well, not quite. It isn't true that there are no factors that affect the rotation of the Earth. There are a variety of forces which influence its rotation, by either slowing it down or speeding it up, such as the tidal force of the Moon's gravity, or the non-solid nature of the Earth's atmosphere and mantle. However, the rotational energy of the Earth (angular momentum) is huge, so small factors like these don't have a great effect on its rotational energy. This is why one day is always 24 hours, and it can't so easily be changed to 23 or 25 hours. It would take a tremendous force to change the rotational speed of the Earth. A spinning top is small and has low angular momentum, so its rotation can be slowed by even dust, and its ability to spin is easily weakened.

Rotating bodies will continue to rotate.

Lesson 19

A Bird in an Airplane

When an airplane is flying (flying level at a fixed speed, or ascending or descending at a fixed speed) there is no difference between the gravity felt by the people inside the airplane and the gravity felt by people on the Earth's surface. So a bird inside an airplane flying at a fixed speed would be able to fly normally as though the airplane were parked at the runway, even if the airplane was flying at 900 km/h (560 mph).

When an airplane accelerates the people inside it feel a force (inertia) in the opposite direction to the acceleration. This is the force you feel pressing you back into the seat when taking off. Likewise, when it decelerates you feel an opposite, forward force. If the airplane climbs sharply, your body is pressed downwards and you feel heavier. Or if the airplane nosedives, you are pressed upwards and feel like you're floating. So the bird would also feel inertia when the airplane accelerates or decelerates and be confused. A passenger plane, however, doesn't accelerate or decelerate very fast, so in the end it doesn't make that much of a difference.

There is a method of flight called parabolic flight, which is where a jet plane accelerates very fast, climbs steeply and then cuts the engine, causing a parabolic rise and fall in altitude. Using parabolic flight, a state of zero gravity can be sustained inside an airplane for up to 20 seconds. So what would happen if we released the bird under a state of zero gravity? The bird would naturally flap its wings and try to fly, but instead of flying, it would start to do back-flips, spinning around and around. On the Earth, birds flap their wings to attain lift, overcome gravity and fly. When a bird tries to do the same thing by flapping its wings in zero gravity, there is no force pulling it down (gravity), so there is a constant lift force pushing upwards on the bird, causing it to spin.

Without acceleration or deceleration, traveling inside something feels just as though it is stopped.

Lesson 20

TELEPORTATION

With the growth of civilization many methods of transport have developed. We all recognize increasingly complex transportation methods such as the horse drawn carriage, the steam locomotive, the automobile, the electric train, the airplane, and the rocket. But surely the ultimate imaginable by humans is teleportation. Wouldn't it be wonderful to be able to see distant relatives, friends or sweethearts immediately?

Teleportation is well-known in the world of movies and comics, whether it's the transporter beaming people in Star Trek, a wizard apparating in Harry Potter, or the Spaceship Yamato in Star Blazers. Teleportation is often used in epic stories which take place across the vastness of space. Even at light speed, it would take 4 years and 3 months to reach the nearest star to the Solar System (Proxima Centauri). Although the Great Magellan Nebula which the Spaceship Yamato was trying to reach is one of our closest galaxies, it would take 170,000 years to reach it at light speed. This is why in stories which take place between stars and galaxies some sort of teleportation system is needed to cross space.

It has been theorized that teleportation could exist in the real universe using space-time tunnels. Scientists say there could be holes scattered across space-time called wormholes. The idea comes from the general theory of relativity, which says that there should be a gravitational distortion around massive objects. You can visualize it as the dented shape in a stretched-out sheet when you place a heavy ball on it. A wormhole is a hole linking one of these gravitational distortions to another. Teleportation using wormholes is possible in theory, but not in practice. The enormous amount of gravitation needed to produce a wormhole requires a black hole, and if a black hole were to draw spaceships and people in at one end, they would be mashed into tiny pieces and wouldn't be able to return alive.

It may be a fantasy to travel through time and space. But our mind can easily fly around the universe without difficulty.

Lesson 21

The Number of Stars

We all use the word star to mean the points of light in the sky. In its scientific meaning, a star is a heavenly body which emits light, like the Sun.

It is said that with the naked eye you can see a total of around 8,600 stars, ranging from bright magnitude 1 stars to dim magnitude 6 stars. Having said that, in this modern age the only way you'll be able to see dim magnitude 6 stars is to go to a remote region with little street lighting. In a city you can see at best magnitude 3 stars.

If you include all the other stars which we can't see with the naked eye, there are 200,000,000,000 stars in the Milky Way alone. Then consider that there are 100,000,000,000 galaxies like ours in the universe. That means that there are 200,000,000,000 stars multiplied by 100,000,000,000 galaxies, or 20,000,000,000,000,000,000,000 stars in the universe! That would be 20 billion trillion, or in scientific terms, 20 sextillion.

The Sun is one ordinary star amongst all of those.

The Sun is just one of an almost infinite number of stars.

Lesson 22

Up and Down in Space

On the Earth, we know that when a person is standing up, their head is pointing upwards and their feet are pointing downwards. This is thanks to gravity acting downwards from your feet towards the center of the Earth.

But what happens in outer space? Surely there is no up and down in zero gravity. If somebody were to stand in an upright position you could just decide that their head is pointing upwards and their feet are pointing downwards, but all directions would change depending on which way they moved.

So what about inside a spaceship? The inside of a spaceship is zero gravity, so the heads and feet of astronauts could point in any direction. There is no meaning to up and down in this sort of environment. But for those who permanently live in space, it would be very useful to have a common rule for up and down. Inside a space station there are floors, ceilings and walls, just like a building on Earth (although any surface could be the floor as it's zero gravity). So, it is decided that the space station moves so that the floor always points towards the Earth, and by fixing the position of the floor, all positions can be decided.

If one direction is decided, all are decided.

* For spaceships and airplanes, we call upwards "zenith", downwards "nadir", forwards "fore", backwards "aft", right "starboard", and left "port".

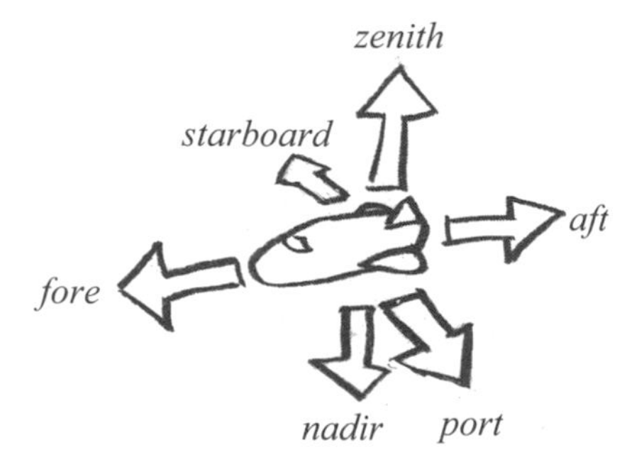

Lesson 23

HOW FAR TO THE STARS?

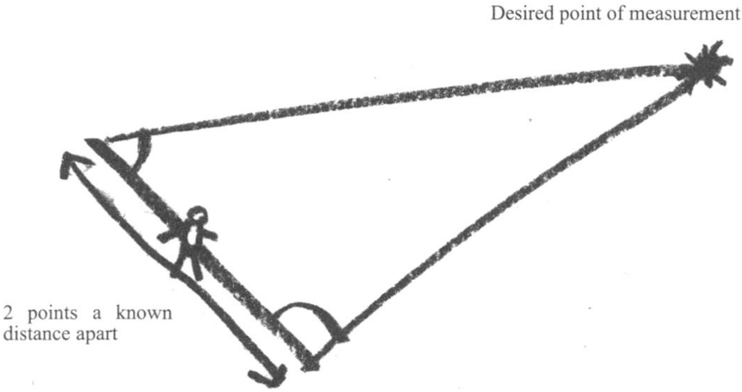

Desired point of measurement

2 points a known distance apart

If you know the angles, you can work out the distance.

Stand somewhere with a view into the distance and hold up your index finger. Then close your left eye. Then open your left eye and close your right eye. You'll notice that the distant scenery doesn't change, but your finger moves position depending on which eye you're looking with. Human and animal eyes are separated a small distance, giving a different viewing angle for each eye. This allows us to tell how far away something is, and is called the principle of triangulation. This difference in the angle when we look from two different locations gives us the distance to an object.

When we use this method of triangulation to find out the distance from the Earth to a star, the distance between the two "eyes", or observatories, has to be as wide as possible.

When you walk down a street at night it feels as though the Moon is walking with you. The reason is because the distance you have walked is tiny compared to the distance to the Moon, so the angle to the Moon is roughly the same even after a few hundred yards. This is even truer for stars in the night sky which are much further away than the Moon. It is easiest to measure the distance to a star using very distant observatories, such as Japan and the USA, or Chile and Australia in the southern hemisphere. But even using these, there are stars that are so distant that we don't have observatories far apart enough to measure it. The Earth just isn't big enough.

The trick that astronomers use to measure the distance to stars more accurately is to make an observation when the Earth is on one side of the Sun, and then another observation on the other side of the Sun, as we orbit around. By comparing the angle to a star in January to that in July, half a year later, you can measure the distance to the star using triangulation.

To judge distance we need a pair of eyes.

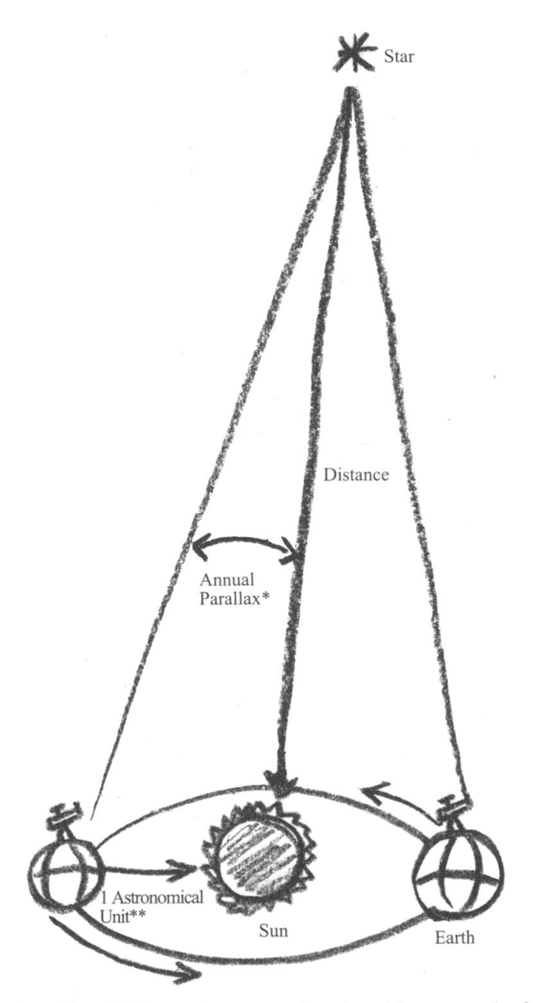

* Annual Parallax: Half the angle to a star when viewed from two ends of an orbit of the Earth around the Sun. The smaller the Annual Parallax, the further away the star.

** Astronomical Unit: The average distance from the Sun to the Earth (149,600,000 km = 93,000,000 miles)

Lesson 24

Temperature in Space

The Sun is the heat source for all hot and cold in the Solar System.

The coldest temperature in the universe is called absolute zero (around -460 ° F), and nothing can be colder than that. How hot an object is above absolute zero depends on how much heat it receives from a star. When light from a star falls on an object its temperature rises and when it doesn't it drops.

The reason the Earth has mild temperatures, ranging from only the minus tens in cold regions to over 100 ° F in hot regions, is because it has a thick atmosphere which allows heat to circulate. By contrast, the Moon has no atmosphere and the ground gradually heats up during the day to around 230 ° F, and at night cools to around -240 ° F. The surface temperature on Venus can be as high as 880 ° F because it is closer to the Sun, and because it has a thick carbon dioxide atmosphere (with 90 times the pressure of the Earth's) which causes a powerful greenhouse effect.

Spaceships or space stations, as well as astronauts performing extra-vehicular activities (spacewalks), are bathed in direct sunlight, so where they face the Sun the temperature rises and where they face away the temperature drops. The exposed areas can reach above 220 ° F, and the unexposed areas below -150 ° F.

This is why space suits for extra-vehicular activities have 13 layers, to protect the body from the extreme temperature variations in the space environment. Thin tubes about the size of spaghetti are attached to the inside of space suits, which pump coolant from the back-pack life-support system around the suit to maintain body temperature. Inside the space station the temperature is controlled to a comfortable temperature for daily work (around 75–80 ° F) so that no space suit is required and astronauts can wear t-shirts and shorts.

Air and water circulation keep temperatures mild.

Lesson 25

Can you Hear Sound in Space?

The sounds we hear every day are air vibrations. We hear sound because waves of energy (sound waves) travel through the air and arrive at our ears.

Sound can also be transmitted through objects other than air. You have probably used a tin can telephone where the sound is transmitted through a piece of string. Or if you put your ear to the table and knock on it you can hear the sound directly transmitted through the table. Sound can even be transmitted as waves through water!

But what about in space? Well space is a vacuum: an area with absolutely nothing in it. That's why sound can't travel through outer space. The surfaces of the Moon and Mercury, which have no atmospheres, are silent worlds.

And what about inside the space station? Inside the International Space Station, there is an atmosphere (nitrogen and oxygen) just like on the Earth. That means the astronauts can breathe normally and sound can travel normally.

If there is something to vibrate, sound can travel.

Lesson 26

Is there Wind in Space?

Wind is air flow, so there can be no wind in the vacuum of outer space. There is also no wind on the surface of the Moon or Mercury.

But on planets with an atmosphere, air flow in that atmosphere will create wind. On Mars there is a thin atmosphere and strong winds blow when the seasons change, stirring up sand storms and tornados. Saturn's 6th Moon, Titan, is also known to have a thick atmosphere. In 2005, the ESA (European Space Agency) Huygens Probe landed on the surface of Titan. You can listen to an audio recording of Titan's wind on the ESA homepage. Imagine that: the sound of wind from another planet. Wouldn't you like to hear that?

Although there is no wind in outer space, there is a kind of wind coming from the Sun called the solar wind. The solar wind is not air flow, but a flow of charged particles (like protons and electrons), called plasma.

When on the Earth, a compass's magnetic north points northward and its magnetic south points southward. Because opposite poles attract and like poles repel, the Earth's geographic North Pole is actually its magnetic south pole and vice versa for its geographic South Pole which is its magnetic north pole. The Earth is a giant magnet that produces a protective magnetic field (an area influenced by magnetism) which surrounds the Earth. When the solar wind is very active, it affects this magnetic field causing a beautiful aurora in the night sky and causing problems for satellites and telecommunications equipment.

People living on the space station need to protect their bodies from the radiation which comes with the solar wind. A space weather forecast predicting solar wind is quite important to people living in space. You can find space weather forecasts on the homepages of telecommunications research organizations.

Planets with an atmosphere have wind.

Lesson 27

A Zero-Gravity World

Inside the space station air is provided at a pressure of 1 atmosphere, the same as on the surface of the Earth. The difference is that inside the space station there is zero gravity. When you put something down on a table inside the zero-gravity space station, it lifts into the air because of tiny vibrations and air flow from ventilation ducts. This is why tape, clips, elastic bands and magnets are widely used inside the space station.

What would happen if gas were emitted inside the space station? Gas (which is a fluid) has a property, it diffuses. Diffusion is a property whereby a substance with varying density mixes to even out the density, and can be seen in the way smoke expands. Diffusion happens in zero gravity as well. A gas emitted inside the space station would mix due to diffusion. So, smells can travel in zero gravity because substances which produce a smell do so by producing small amounts of gas.

One of the properties of gas or fluid is convection. Convection is where a hot gas or liquid becomes lighter and rises, causing the gas or liquid above to sink. It's the same process that happens in a bath making the hot water rise to the top and the cold water sink. You might be able to guess that convection occurs as a result of gravity, so it doesn't occur in zero gravity. When you light a candle on Earth we all know it produces a tall, elongated flame, which is caused by airflow due to convection. But because there is no convection inside the space station, the flame would not stretch upwards at all. It would be round.

If one were to spray a little water inside the space station, what would happen? Water has a property called surface tension which always tries to keep the surface area to a minimum. On the Earth you can see this effect when you fill a glass right to the top and the water bulges slightly above the rim of the glass. In space there is no pressure keeping the water in the glass, so it floats away! Then that same property of surface tension forms the water into a ball. These small water drops form into spheres and float around freely.

In zero gravity, a flame on a candle is round.

Lesson 28

THE STRAIGHT LINE AND THE SHORTEST PATH

A straight line is a line which does not bend, and is the shortest path connecting any two points.

In our everyday lives this is true, but according to the theory of relativity an interesting thing happens which makes a straight line not a straight line to its observer.

Light travels by the shortest path, or a straight line. But light which passes too close to a star or other massive object has actually been seen to bend because of gravity. This is proof of Einstein's General Theory of Relativity.

For simplicity let's take a 2 dimensional example. If you look at a map of the earth, we know that we can find the shortest distance between 2 points by drawing a straight line between them with a ruler. However that assumes that the ground is flat. In practice there are high points and low points on the Earth, high mountains and deep lakes. Is the shortest distance between 2 points, even if they are separated by a giant obstacle (such as Mt. Fuji), really a straight line?

In reality of course there is likely to be a path winding around the peak of the mountain which would be the shortest path. So although we say a straight line is the shortest path, in this case the shortest path is not straight, but it bends. If there was no mountain there, the shortest path would be a straight line between the 2 points. But even that line is a curve because the Earth is spherical.

In the same way that land can be bumpy or be a curve which appears flat, the universe is actually warped by gravity. If you go close to a massive star the shortest distance becomes a curve. A straight line is no longer a straight line.

Light travels on the shortest path, but this isn't always a straight line.

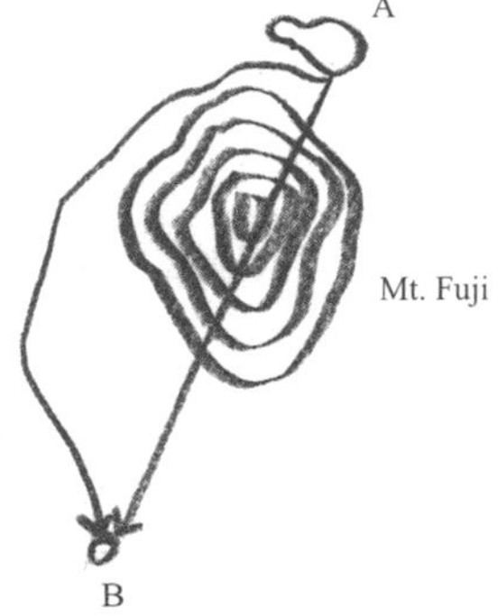

Lesson 29

Bodily Changes in Space

The longest continuous stay in space was 437 days by a Russian cosmonaut named Poljakov on space station MIR, but even a short 2 week flight on the Space Shuttle will result in some bodily changes.

About 70% of the human body is water. When we're standing on the surface of the Earth, gravity pulls our bodily fluids (blood and so on) down to our feet, but even so our bodies have learned to push bodily fluids up to our brains. When you go up into space there is no gravity and no force pulling your bodily fluids down, so they tend to move towards the upper half of your body. Pretty soon your face begins to swell and you start to lose concentration. Astronauts call this condition *Moon face* because your face swells up into a round Moon shape. It's exactly the same thing that happens when you stand on your head and the blood rushes to your face. However once you've been up in space for a few days the condition improves, as the body begins to adapt to zero gravity.

Another amazing thing is that when you go into space you get taller. People gain on average about one inch. Some people, who really "grow," gain two or three inches. This is because the cartilage between your bones expands when the pressure of gravity disappears. Of course, you would return to your normal height once you return to Earth.

One problem with staying in space for a long period is the weakening of bones and muscles. In zero gravity your body doesn't bear any weight and the only movement you usually make is a simple push against a wall. When your bones don't bear any weight they lose calcium and become hollow, and your muscles become weaker. This wouldn't be a problem if you were to spend your whole life in space, but if you were ever to return to Earth you wouldn't be strong enough to move normally.

People living on the space station actually use a stationary bicycle and a running machine for two hours each day to provide a force on the body and do physical training to prevent their bones and muscles from wasting away.

Exercise is compulsory in space.

Lesson 30

Objects in Space

In 1957 the former Soviet Union was the first country to launch a man-made satellite, called Sputnik 1. Since then about six thousand satellites have been launched and there are more than three thousand circling the Earth right now. Satellites and other artificial bodies circling the Earth in relatively low-altitude fly at an altitude between 80 to 2000 km (50 to 1200 miles), which is called LEO, Low Earth Orbit; manned spacecraft such as the space station and space shuttles orbit within an LEO. Satellites circling the Earth at relatively high-altitude fly at an altitude of about 36,000 km (22,000 miles), the height at which geostationary satellites orbit. This is called GEO, Geostationary Orbit. Even though so many satellites fly at the same orbit, the chance of them colliding is very small because the size of these satellites is like an insignificant dot compared to the giant expanse of space*.

There is, however, a lot of garbage floating around in these orbits. It is composed of fragments broken off from satellites and fragments of launching rockets. It is estimated that there are more than 8,000 pieces of this space debris over 10cm in size and more than 30,000,000 pieces over 1mm in size. Collisions of space debris with satellites or the space station is starting to become a problem. Space debris above 10cm in size can be observed using telescopes and radar, so the position of spacecraft (satellites and the space station) can be shifted to avoid collisions. Also, the body of the space station is protected by shielding which can handle collisions with debris up to 1cm in size.

The problem is the space debris between 1 cm and 10 cm. There is no way to predict when such debris will approach a spacecraft. If it hits it will cause damage. The probability of debris bigger than 1cm hitting a spacecraft is very small, but tiny particles strike the body, windows and solar panels of spacecraft quite regularly.

The Space Control Center, part of United States Strategic Command, tracks space debris larger than 10 cm and protect spacecraft against it. There is also a facility to track space debris in Japan called the Bisei Spaceguard Center. An optical telescope observes space from Bisei, acting as a space security guard that prevents collisions.

There is garbage even in space.

* On February 10, 2009, U.S. and Russian communication satellites crashed in space. It was the first accidental collision occurred between two intact satellites in space. It was monitored that about 600 pieces of space debris which is more than 10 cm in size were left after the crash and could stay in orbit for up to 10,000 years.

Lesson 31

Direction of Rotation in Space

In the film *Superman* there is a scene where Superman flies around the Earth causing it to rotate backwards, which makes time flow backwards so he can bring Lois Lane back to life.

But what would really happen if the Earth were to spin backwards? Well, the Sun would rise in the west and set in the east. Constellations would circle the Pole Star clockwise rather than counter-clockwise. Westerly winds would not blow from the west, but the east, and hurricanes which had previously rotated counter-clockwise in the northern-hemisphere would start to rotate clockwise. There would also be big changes to ocean currents.

Time, though, would not reverse.

In reality rotating the *Earth* backwards wouldn't cause t*ime* to move backwards, but if some supernatural force were to make *time* move backwards then the *Earth* would also rotate backwards.

Thus by this line of reasoning, it must be that Superman reversed time rather than the rotation of the Earth.

The Earth rotates towards the east (counter-clockwise if viewed from above the North Pole). Planets including Mercury, Mars and Jupiter also rotate to the east. Venus is the only planet in the Solar System which rotates to the west very slowly (1 rotation takes 243 days). It seems that a brake is put on Venus' rotation by its thick atmosphere and its axis of rotation has inverted.

Time cannot go backwards

Lesson 32

Space and Garbage

Two hundred and twenty million tons of garbage (not including industrial waste) is produced in America each year. This amount could fill the Houston Astrodome 600 times. The only way to deal with this garbage is to incinerate it or bury it in landfill. The shortage of landfill sites is now a problem in cities the world over, and you have to worry whether one day the world will be filled with garbage.

There are people considering disposing of garbage somewhere in the vastness of space. A particularly elegant solution might be to send garbage into the Sun. The temperature of an incinerator is around 1,500 to 1,700 ° F, but the surface of the Sun is 10,000 ° F (at that temperature even iron and lead would melt and evaporate). The center of the Sun has an ultra-high temperature around 27,000,000 ° F, and ultra-high pressure, too.

The Sun is 109 times larger than the Earth in diameter. It has an enormous volume, 1.3 million times that of the Earth, so it could easily swallow the whole Earth, never mind a few scraps of garbage. But launching garbage into space or into the Sun is not practical. Firstly, it is very expensive to launch a rocket. It takes around $100 million to launch a single rocket. Despite that, we can send objects up to four tons into a geostationary orbit, although the amount that we can send beyond that orbit (for example to the Moon and other planets) is smaller. Another problem is that the rocket itself weighs 300 tons. It would be crazy to turn a 300 ton rocket into garbage just to throw away four tons of trash.

So what about garbage generated while living in space? Astronauts now spend long periods living on the space station and generate all sorts of garbage, such as food containers. Lavatory waste also has to be disposed of. Disposal of these things is done using an unmanned spacecraft called Progress which regularly (once every few months) transports food and necessities to the space station. After unloading all the materials transported from Earth, the spacecraft is stuffed with garbage. It then detaches and is incinerated by the burning heat when re-entering the atmosphere, just like a shooting star. Shooting stars are natural objects in space, called meteoroids, that burn and light up when entering the atmosphere. Next time you see a shooting star, you might wonder whether it's actually just the space station's garbage.

Earth's garbage can't be sent into space.

Lesson 33

The Earth's Atmosphere

The law of gravitation says an object with weight and another object with weight (an object with mass and another object with mass) are pulled towards each other. We normally don't think of air as having weight, but it does. Just like when you jump into a pool and don't feel the weight of the water, we don't feel the weight of air because we are inside it. The air on the Earth is held around the Earth by the force of gravitation between the Earth and the air, or in other words it is pulled down by the Earth's gravity.

We call air that surrounds planets an atmosphere. The atmospheres on other planets have a different composition than the Earth's, but even Venus, Mars and Jupiter have atmospheres. The Moon and Mercury have no atmosphere because they are too small and light, meaning they have weak gravity, and they can't hold air around the surface.

Planets with strong gravity can trap air, but no air can be held on planets with weak gravity or in outer space. This is why there is no air in outer space, only a vacuum.

Even air is pulled down by gravity.

Lesson 34

A State of Vacuum

A vacuum is a volume of space that is void of any matter. Even though it looks like the space around us is empty, it still contains air. A vacuum is a volume without even that. There is a machine called a vacuum pump which sucks up air and reduces the pressure in a container. A vacuum is created by removing all the air.

Having said that, it is impossible to create a perfect vacuum, no matter how efficient the vacuum pump. There will always be a small number of air molecules left. If you go into space you can have a better vacuum than you can create on Earth. But even in space some atoms will remain and it will not be a perfect vacuum. We call this an extreme high vacuum.

So what can happen in a vacuum state? Well, if you lower the pressure in a container and approach a vacuum, liquids will boil even at lower temperatures. Water rapidly boils and evaporates even at room temperature. If a human were to go out into outer space without a spacesuit their bodily fluids would boil and evaporate, and they would completely dry out like a mummy. Also no sound can travel through a vacuum. Sound is transmitted vibrating air and materials. So a vacuum, which is space without anything, is a world without sound. Astronauts on a space walk communicate with astronauts in the space station via communications devices. Unlike sound, electromagnetic waves do travel in a vacuum.

Vacuum packed food is food which has had the air removed from the package using a vacuum pump, to prevent the food from spoiling due to oxygen. It isn't a perfect vacuum, but it keeps the food fresh and allows it to be stored for long periods.

In space, extreme high vacuums can be achieved which can't be created on Earth.

Lesson 35

GETTING INTO SPACE

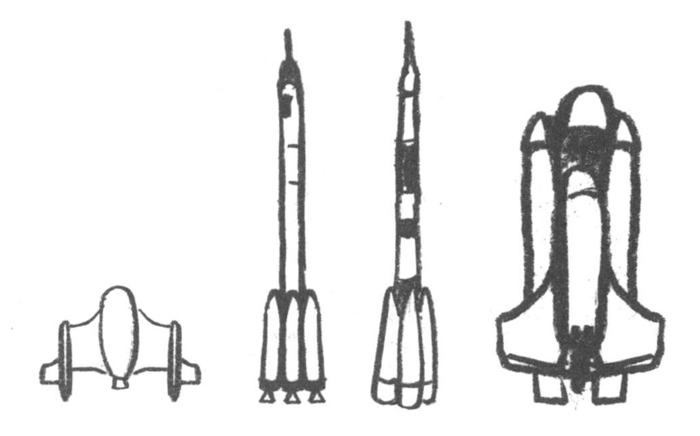

There are currently 4 types of manned spacecraft capable of taking humans into space.

Space Shuttle (USA)

Soyuz Spacecraft (Russia)

Shenzhou Spacecraft (China)

SpaceShipOne (USA / Civil)

The Space Shuttle is a re-usable rocket and spacecraft with wings like an airplane. When it returns to Earth it glides through the atmosphere. The Space Shuttle is the only spacecraft which can take up as many as 8 passengers per flight. Since the maiden flight of the Colombia in 1981, it has launched more than 120 times.

The Soyuz spacecraft is a spaceship for 3 passengers which has been in use and continual development since 1967. Passengers are accommodated in the tip of a giant 165-ft tall rocket, and only the top 8-ft diameter capsule returns to Earth, floating down by parachute.

The Shenzhou spacecraft was first launched in 2003. First, numbers 1 through 4 were launched unmanned with dummy astronauts. Then the Shenzhou-5 launched in 2003 carrying a single astronaut, and safely returned after orbiting the Earth 14 times in 21 hours. In achieving this China became the third country after Russia (former Soviet Union) and the USA to send a person into space successfully. China now has its own space development program.

SpaceShipOne is the first civilian spacecraft. It was successfully launched in 2004 by the American company Scaled Composites. It is flown up to a height of 15 km (9 miles) hanging under a jet airplane, and powers into space using its rocket engine. SpaceShipOne will be used as a spacecraft to take ordinary people into space.

These are the 4 current methods for getting into space. Which one would you like to take?

There are 4 types of ships which take you into space.

Lesson 36

The Theory of Rocket Flight

Have you ever thought about the difference between the flight of an airplane and a spacecraft?

An airplane flies within an atmosphere. Both jet and propeller airplanes move by taking air in at the front and pushing it out of the back. An upwards force (called lift) is provided by the resistance of the wings moving through the air. The principle of airplane flight is similar to swimming the front crawl.

A rocket, however, works by burning (and exploding) fuel inside itself, forcing it out of a nozzle (thrusters), and producing lifting power from the reactionary force. This force comes from what is called the action-reaction law, or Newton's third law. Imagine you're wearing roller skates or ice skates (to reduce the friction your feet create normally with the ground) and you make a quick pass with a basketball. Your body would be moved backwards by the reactionary force. The force of you passing the ball is the action force, and the force you feel pushing you backwards is the reaction force. You can experiment with a balloon. Blow up a party balloon without tying it, then hold it in the air and let go. It is the reaction force from the released air which makes it fly around the room. The air leaving is the action force and the opposite force which moves the balloon is the reaction force. Because rockets use this action-reaction law, they can fly both in an atmosphere and in a vacuum.

If you've ever seen a video of a rocket launch, you may imagine the fuel coming out of the main engine thrusters is pushing against the floor to enable the rocket to take off. When the rocket is airborne, too, you might say the fuel coming out is pushing against the surrounding air. However, even if neither the floor nor the surrounding air was there and the rocket was flying in the vacuum of space, the fuel would still create a reaction force.

So we know rockets can fly without air, but don't rockets need oxygen to burn fuel?
Well, the Space Shuttle uses liquid hydrogen for fuel, but hydrogen can't burn alone. The reason hydrogen explodes when you light it is because there is oxygen in the air. That's why the Space Shuttle also stores liquid oxygen in its fuel tanks to help the hydrogen to burn, enabling it to fly in a vacuum.

Rockets fly thanks to the action-reaction law.

Lesson 37
The International Space Station

Did you know that right now there is a space station flying over our heads with people on board? We are already in the era of life in space.

The International Space Station began orbiting at 400 km (250 miles) above the Earth in 1998 with the cooperation of 15 countries, and since 2000 has had a permanent crew of 2 to 3 astronauts on 6-month shifts. It was transported into orbit by more than 40 launches by the Space Shuttle and Russian rockets, and assembled by astronauts using space walks and robot arms. When complete it will be the size of a football field. It includes modules for living, storage, research, and docking. The most recent research module, Kibo, was made by Japan and was installed in 2008 and 2009.

Lots of experiments and research can be carried out on board the space station. New materials which can't be created on Earth are being developed due to its zero gravity environment, and research is being done via biological and medical experiments on how the environment affects people living there.

The space station is the borderless frontier of the human race. It makes you wonder what sort of place the space station is and what sort of life the occupants have. The space station flies at around 400 km above the Earth at a speed of 7.8 km (4.8 km) per second. This is an incredible speed. It is equivalent to 28,000 km (17,000 miles) per hour, or 23 times the speed of sound on Earth (Mach 23). If you could run at that speed you would be able to finish a marathon (26.22 miles) in 5 seconds! Or you could travel from San Francisco to Los Angeles in about 1 minute and 10 seconds!

The space station is a new place for humans to live.

Lesson 38

Life on the Space Station – Language and Food

The space station is jointly built and run by 15 countries. Those 15 countries are the USA, Canada, Europe (Belgium, Denmark, France, Germany, Italy, the Netherlands, Norway, Spain, Sweden, Switzerland, and the UK), Japan, and Russia. A Brazilian astronaut and a civilian from South Africa have also been to the space station. The space station is a borderless world where people of all nations live together. In this kind of situation an official language has to be decided.

It has been internationally agreed that English is the spoken language of the International Space Station. The French, Germans, Russians, and Japanese people all work using English inside the International Space Station. However even though the official language is English, Russian technology is used almost as often as American technology, so knowledge of the Russian language is just as important as English for a smooth life on board the space station. Astronauts of all nations not only have to brush up their English skills, but also receive regular practice in Russian.

Astronauts boarding the space station will spend around 6 months in space, so the food they will eat over this time (space food) has to be easy to cook and have a long shelf life. Astronauts can only eat fresh food such as fresh fruit and vegetables during the first few days after their arrival or after a supply ship delivery. Their everyday food is either freeze-dried or boil-in-the-bag. You probably know freeze-dried foods in the form of cup noodles, which are frozen and dried in a vacuum. Boil-in-the-bag foods are sterilized at high temperatures and pressures and then sealed in an airtight bag, and are often used for easy-cook curries. Drinks are carried into space in airtight silver packs. They are then mixed with cold or hot water in space and drunk through a straw. Drinks can't be drunk from a cup because the space station is zero gravity and if the container isn't airtight the drink would just float around in the air. Salt and pepper would also float around in the air getting into people's eyes, so they are dissolved in water and taken up in plastic bottles. They can be added to food in the same way you put drops in your eyes using a dropper.

Space food in the space station is made in the USA and Russia. They say there are around 200 to 300 varieties including scrambled eggs, potato gratin, teriyaki chicken, mushroom soup, chicken noodles, and smoked turkey. The most popular meal with NASA astronauts is shrimp cocktail. It's boiled shrimp in chili sauce. It is often eaten as an appetizer on Earth. You might say its spicy taste is the best in the universe. Many astronauts that have been into space say they preferred spicy foods over their normal preferences while they were up there.

If you live in space, you may come to love spicy foods.

Lesson 39

Viewing the Space Station

Did you know it's possible to see the space station with the naked eye from the surface of the Earth? The International Space Station flies at 400 km (250 miles) above the surface of the Earth. You can see light from the Sun reflecting off the space station when it's bright in the altitude and dark on the ground, i.e. at just before dawn and just after twilight.

The space station revolves around the Earth once every 90 minutes with an orbital inclination of 52 degrees. In other words, this means the space station flies over the northern hemisphere in 45 minutes, crosses the equator, flies over the southern hemisphere in 45 minutes and crosses the equator again, and so on and so forth. Because during that 90 minute interval the Earth rotates by 22.5 degrees around its axis, the space station's position in the sky is constantly shifted by 22.5 degrees each trip around. The position of the station as seen from the ground changes day by day. You can see the space station above your location, in good conditions, about once a month. When it is visible at dawn one month, it will be visible at dusk the next. So, if you want to see it you may have to wait a month or two. You can find out where to look on NASA's official website. The orbit of the space station is corrected, so the exact location and time for observation may be changed. Check the website often and keep an eye out for a prime viewing opportunity where you live.

The space station does fly across the night sky, but not like an airplane does. While an airplane makes a sound and has flashing lights, the space station silently moves along its orbital track with a fixed speed. In very good conditions the brightness of the space station can be as high as magnitude minus 1, about as bright as the brightest stars. It shines because it reflects light from the Sun, in the same way as the Moon and planets, such as Venus and Jupiter. Although the space station looks like a moving star, it is completely different from a shooting star. A shooting star can only be seen for a fraction of a second as it falls, but the space station glides across the night sky for 2 to 3 minutes. You might call the space station a man-made star. You can also see artificial satellites in the night sky (around magnitude 3), but the space station is far, far brighter. The main difference with satellites is that there are people living on board that star-like twinkle in the sky, otherwise known as the International Space Station, right now. It is an amazing feeling to look up with that in mind.

Within that star-like light are astronauts working, eating, exercising, and gazing down at the Earth. When we see Venus at dawn we call it the Morning Star, and so might the space station be a signal of a new dawn for mankind.

You might call the space station a man-made star.

Lesson 40
Space Exploration Spinoffs

New technologies are constantly needed to overcome the challenges of mankind's advance into space. This gives rise to ever more technological advancements which are useful in all sorts of areas of our everyday lives. These are called technology transfers, or spinoffs.

One example is a material for space suit gloves which stays warm when it's cold and cool when it's warm. It's called Outlast®.

Products made from Outlast, e.g. shirts, underwear, socks, shoes, gloves, pillows, comforters, and futons are sold in more than 60 countries worldwide. Hang gliders were invented based on a spacecraft recovery system. Basketball shoes which boost jumping ability and reduce foot impact were developed from technologies used in spacesuits. The communication system used between astronauts was adapted for use in television remote controls and cordless appliances. Technology for recycling water within spaceships was adapted into technology to purify rivers and ponds. Laser technology used in lighting and medical equipment was originally developed to accurately measure the distance between the Earth and the Moon for the Apollo Project.

Another well-known example is Velcro®. This was invented by George de Mestral in Switzerland in 1948. The invention happened when Mestral was out hunting deep in the mountains with his dog. He noticed his clothing and the dog's fur had picked up countless burrs, and he had a good idea. The burrs use the mechanism of attaching themselves to the fur of animals to spread their seed over long distances. Mestral used this concept to invent Velcro, which can be attached and removed many times. Mestral thought of commercial applications for Velcro. He produced a pair of pants which use Velcro as a zipper substitute, and put a large amount of them on the market. However, he didn't sell any of them. It seems it just wasn't stylish enough. Mestral had built up large quantities of unsold stock and was at a loss, until he hit upon the idea that they can be used in space. In space there is zero gravity so objects can freely float around. It had occurred to him that it would be incredibly useful to have a device which would allow you to fix and free objects at will. So, Mestral took the idea to NASA, where it was adopted and used to fix covers to satellites and to hold objects inside the space station. Space technology is now used in everyday life and in all manner of ways.

Developments of space technology make our everyday life more convenient.

Lesson 41
COMETS

Comets are astronomical objects which sometimes approach the Sun in an elliptical orbit.

Haley's comet approaches the Sun every 76 years, the Hyakutake comet every 17,000 years, and the Hale-Bopp comet every 2,500 years. Comets travel through dark space for decades or even millennia and turn bright only for a few months when they approach the Sun and are close to the Earth. As a comet approaches the Sun, its core is heated by the Sun and it emits gas and dust. This gas and dust is then blown into a long, shining tail by the Solar Wind.

A comet lights up as it approaches the Sun.

Lesson 42

The True Nature of Shooting Stars

Shooting stars (or meteors) occur because fine dust drifting in space (from under a millimeter to centimeters in size) is drawn in by Earth's gravity, enters the atmosphere, and produces light due to collision with the air.

This dust is mainly the dust that is left behind in the paths of comets. As the Earth revolves around the Sun, it passes through certain areas where there is a lot of dust and these are the positions where we can see lots of meteors. We call these meteor showers.

People sometimes make a wish on a shooting star, but you better wish fast because a single shooting star lasts for less than a second. As soon as you notice it, it has gone. The most active time of year for meteor showers is the Perseids meteor shower from the 12th to the 13th of August. If you go to a dark region away from big cities, you may be able to see dozens of meteors over the course of an hour. If it's the time of Perseids, it might be a little bit easier to make a wish on a star.

Shooting stars are small amounts of burning dust.

Lesson 43

METEORITE

A meteorite is an object which was floating in space, got pulled in by the Earth's gravity, and landed on the ground without completely burning up. You could also describe them as large shooting stars which have reached the ground without burning up.

The largest existing meteorite is the Hoba Meteorite in Namibia, Africa, which is nine feet wide at the widest point and weighs 60 tons. But what would happen if an even heavier meteorite were to fall on the Earth?

Checkout Receipt
Willard Library
Renew your items online:
http://catalog.willardlibrary.org
Title: The World's Easiest
Astronomy Book
Call Number: 520 NAK
Item ID: 33330007998190
Date Due: 7/19/2017

Around 65 million years ago, a huge meteorite of 10 km (6 miles) in diameter fell on the Yucatan Peninsula in Mexico. The crater is a massive 170 km (110 miles) across and the rock fragments and dust thrown up by the shock of the impact clouded the Earth's atmosphere. This meant that light from the Sun couldn't reach the ground enough and global temperatures dropped. This sudden climate change and environmental change are thought to have caused the extinction of the dinosaurs.

If that meteorite hadn't struck the Earth 65 million years ago, perhaps the evolution of life on Earth would have been quite different. The dinosaurs might not have died out, and might have evolved in all their splendor, gained intelligence, and may even have ruled the world today. Humankind might have been merely rat-like creatures running from under the dinosaurs' feet, and it would be the dinosaurs reading this book!

A giant meteorite has the power to change the fate of the Earth.

A meteorite changed the fate of the Earth.

Lesson 44

Why Go Into Space

The Earth has existed for 4.6 billion years, and life emerged 4 billion years ago in the primordial seas on the earth. For eons, life continued to evolve in the warm and mild environment of the sea. Life evolved from single-celled organisms, to multi-celled organisms, to complex aquatic life forms. Around three to four hundred million years ago, life moved from the seas to the land. The land was a harsh environment, with temperature and humidity changes, winds and storms, and where ultra-violet rays and various kinds of radiation rain down from the Sun. So what advantages were there to live on the land when the sea is so warm and mild? Well living on the land forced animals with bones to develop legs, and then four-legged amphibians evolved, followed by reptiles. Then came forearms which allowed them to use tools, intelligence developed, and two-legged mammals evolved, followed eventually by humans.

Then in 1961, Russian astronaut Yuri Gagarin was the first human to go into space. Outer space is a very difficult environment for humans to live in. It is a world with no air to breathe, and where radiation harmful to the body constantly rains down. Indeed you might well ask what advantages there are to exploring space when we can breathe and live normally on the Earth without a life-support system.

I believe it is because we have a desire to seek out new frontiers and expand the boundaries of human existence. Just like the first creatures to move from the sea onto the land, mankind also strives to reach out from the Earth into space. In the far future, the time will surely come when countless people are born, grow up, and spend their whole lives in space. We are truly standing at the doorway to the space age.

The desire to go into space is engraved in our DNA.

Commentary

Kenichiro Mogi

This book is written to make the universe easy to understand. It is written informatively yet comfortably, as though you are simply hearing about something going on in your friend's back yard. One has to possess deep knowledge of the universe to be able to write with such familiarity. Only Mr. Nakagawa with his many years of experience in space science could have achieved this.

Although I was born in 1962, I have always had a nagging doubt about space science. I was only just 6 years old when Apollo 11 first carried men to the surface of the Moon. I remember the excitement of the adults staying up late to watch the satellite transmission. At the time I thought mankind would gradually advance into space from there, but space exploration seemed to develop very little, and somehow our hopes of Moon bases, space stations and space travel faded. I had thought a glittering future had arrived, but mankind's concerns began to look inwards. Of course everyday life and politics are important, but I felt that somehow we had lost some of our hopes and dreams. This was the source of my nagging doubt.

It is wonderful to see that recently, private space development is getting up to speed, that there are now space tourists, and there is at last a resurgence of interest in space. I believe that Mr. Nakagawa's The World's Easiest Astronomy Book will further help bring the wonders of space home to readers.

Indeed, one could say the area of space around the Earth is like our back yard. At 13.7 billion light years across, we simply cannot grasp the vastness of the universe as a whole. But the Moon and the other planets in our Solar System are like our back yard, and our exploration of this back yard universe is sure to be filled with amazing treasures like the existence of other life forms, or hidden clues about the history of the Earth. As Mr. Nakagawa has so persuasively written, the technology to carry out this exploration is in development, so we all have a duty to make that leap.

While it is fun to immerse oneself in a fantasy, the myriad wonders of the universe are on the verge of becoming reality. One can't help but gasp in wonder at the vastness of the universe, and when we are moved in this way it impacts on every aspect of our lives. The universe is right here and right now, and is no less than a glorious fantasy.

KENICHIRO MOGI

Born in Tokyo October 20th, 1962, Neuroscientist, Senior Researcher with Sony Computer Science Laboratories, Inc., Doctor of Physics, Guest Professor at the Tokyo Institute of Technology (Neuroscience, Cognitive Science), Part-Time Lecturer at the Tokyo University of the Arts (Artistic Anatomy).

After graduating from the Department of Science and the Department of Law at the University of Tokyo, he completed a post graduate course in Physics at the University of Tokyo Graduate School. Dr. Mogi is currently succeeding his positions at RIKEN and Cambridge University, researching the relationship between the brain and the mind through the concept of Qualia (the quality felt by the senses). He also engages in literary and artistic criticism.

Appeared in Trends in Professional Employment on NHK (Japan Broadcasting Corporation). Recent works include The Brain and Imagination (Hideo Kobayashi Award / Shinchō-sha), The Advent of Qualia (Bungei-shunjū), Life Inside the Brain (Chūō Shinsho), and Brain Insights (Shinchō Shinsho).

The World's Easiest Astronomy Book

Author **Hitoshi Nakagawa**
Binding and design **Shimpachi Inoue**

Translated by **Jon Lawson**
Arranged by **TranNet KK**

Published by **One Peace Books, inc. ,New York, New York**

Special thanks to **Jun Tagawa**

One Peace Books, inc.
57 GREAT JONES STREET NEW YORK, NY10012 USA
TEL: 212-260-4400 FAX: 212-995-2969
URL : http://www.onepeacebooks.com

PRINTED IN U.S.A